Ana Beatriz Alves de Araújo
Maria da Conceição Freitas Moura

Agriculture, Science and Technology

Ana Beatriz Alves de Araújo
Maria da Conceição Freitas Moura

Agriculture, Science and Technology

Knowledge and resources to increase productivity

ScienciaScripts

Imprint
Any brand names and product names mentioned in this book are subject to trademark, brand or patent protection and are trademarks or registered trademarks of their respective holders. The use of brand names, product names, common names, trade names, product descriptions etc. even without a particular marking in this work is in no way to be construed to mean that such names may be regarded as unrestricted in respect of trademark and brand protection legislation and could thus be used by anyone.

Cover image: www.ingimage.com

This book is a translation from the original published under ISBN 978-613-9-72497-0.

Publisher:
Sciencia Scripts
is a trademark of
Dodo Books Indian Ocean Ltd. and OmniScriptum S.R.L publishing group

120 High Road, East Finchley, London, N2 9ED, United Kingdom
Str. Armeneasca 28/1, office 1, Chisinau MD-2012, Republic of Moldova, Europe
Printed at: see last page
ISBN: 978-620-8-33569-4

SUMMARY

CHAPTER 1

WATER EROSION ON UNPAVED ROADS AND STRATEGIES FOR CONTROLLING SEDIMENT PRODUCTION

Ana Beatriz Alves de ARAÚJO Agricultural and Environmental Engineer, PhD, UFERSA, Mossoró/RN.

1. INTRODUCTION

Unpaved or side roads make up approximately 80% of the Brazilian road network, with 1.36 million kilometres out of a total of 1.7 million kilometres of roads (DNIT, 2014).These roads are of fundamental importance to the Brazilian economy, as they transport all agricultural production from production centres to final consumers, and are the main way of transporting food. In addition, they provide the rural population with access to basic services such as education, health and labour needs, and play a significant environmental role in conserving soil and water resources (Oliveira et al., 2009).

However, poorly maintained side roads hinder the flow of agricultural production, making it difficult or even impossible to transport products and people. This leads to an increase in transport costs, which are passed on in the price of the final product to the consumer, as well as affecting the quality of life in rural communities (Oliveira et al., 2011).

According to Enriquez et al. (2015), unpaved roads are one of the factors responsible for the formation of sediment, with direct interference in hydrological processes, and the transport of this sediment causes soil erosion, caused mainly by surface runoff.

Water erosion is defined by Morgan (2005) as the action of soil particles being washed away and carried away by rainwater, mainly through surface runoff. This degrading process has direct consequences for traffic and the surrounding areas, such as soil loss and compromised roads, as well as carrying sediment into water resources, affecting their quality and possible multiple uses.

The development of prediction models is important for choosing an effective technique for containing erosion on unpaved roads, with consequent conservation of water resources, since soil conservation is closely linked to water conservation, and thus the models developed so far have been aimed at water quality.Several studies on water erosion on unpaved roads can be found in the literature, but they are largely concentrated on the European continent. The scarcity of publications on the conditions and conservation practices applied in Brazil, coupled with the economic, social, cultural and environmental importance of these roads for the country, means that more studies like this one are essential to diagnose the condition of side roads and guide future actions to conserve and improve them.In view of the above, the aim of this study is to present a literature review on the occurrence of water erosion on unpaved roads and to present strategies for controlling sediment production and transport.

2. THEORETICAL FRAMEWORK

2.1 Road network in Brazil

Developing countries like Brazil have many unpaved roads. According to 2014 data from the National Road System (SNV), Brazil has 1.7 million kilometres of roads in its territory, of which 1,363,740 kilometres are side roads, making up 79.5%, and the percentage of paved roads is just 12.9%, with 221,820 kilometres (DNIT, 2014).

Brazilian roads were opened without any kind of planning, basically following the location of rural properties and the characteristics of the terrain, such as soil type and relief which, in extreme climatic events such as torrential rains, facilitated the evolution of erosion processes on the roadbed and its banks, in such a way as to jeopardise its use and promote surface runoff, also causing environmental problems by carrying sediment to water bodies (Casarin and Oliveira, 2009).

The absence of a surface run-off drainage system is one of the causes of damage to unpaved roads, causing water to be dammed up in the roadbed and its verges or to run off above the surface of the road. of the conditions that the soil can withstand. These problems occur due to the lack of adequate design and structures for these roads, causing and aggravating erosion processes on the roads and their margins, as well as difficulties in vehicle traffic (Oliveira et al., 2011).

Several authors list important factors for erosion control on unpaved roads, such as correct estimation of the drainage system, knowledge of the soil's capacity to restrict surface runoff, the adoption of mechanical containment practices, erodibility, critical soil shear stress and soil erosion rate (Griebeler et al., 2005; Casarin and Oliveira, 2009; Enriquez et al., 2015).

2.2 Water erosion on unpaved roads

According to Rotta (2012), erosion is conceptualised as a mechanical process that involves the disintegration, transport and deposition of soil particles. Morgan (2005) presents some erosive agents that contribute to the detachment of solid particles, such as water, ice, wind, temperature variations, organisms (plants and animals) and humans.

Koetz et al. (2009), on the other hand, discusses two forms of erosion: natural or geological erosion, and accelerated erosion. Natural erosion takes place under balanced conditions for soil formation and is a slow, constructive process that does not negatively affect the environment. Accelerated erosion, on the other hand, occurs due to anthropogenic actions, in which erosion rates exceed those of soil formation, consequently leading to soil degradation.

Water erosion is the main erosive agent on unpaved roads. Silva (2011) subdivides water erosion into four phases: impact, disintegration, transport and deposition. In situations where the intensity of rainfall exceeds the infiltration capacity of the soil, surface runoff is produced, which in addition to dragging the materials disaggregated by the impact of raindrops, also causes the detachment of soil particles. According to Liu et al. (2014), this detachment capacity by surface runoff is proportional to the increase in runoff depth and slope gradient.Rotta (2012) presents the process of water erosion in two forms, laminar - which is the diffuse flow of water, resulting in the progressive removal of water. linear - caused by concentrations of surface water flow in small channels, which can be of the furrow, gully or gully type - which depends on the surface conditions and the way in which the erosive agent acts. Similarly, Carvalho et al. (2009) defines water erosion, but the author refers to the former as entrenchment erosion and the latter as furrow erosion.As sediment is transported to water sources, the water storage capacity of watercourses is reduced, which consequently leads to a series of problems, such as frequent flooding events, an imbalance in the dissolved oxygen balance in the water, increased turbidity, higher water treatment costs, an increase in the cost of draining the reservoir and, above all, structural damage to homes close to the water resource (Silva, 2011).

Side roads or unpaved roads are understood to be geographical elements present in rural landscapes, which allow regions to be interconnected. According to Van Meerveld (2014), they can originate from a path that has developed over time, or be designed and built according to a geometric project, cross-sections and a drainage system, with the layer being the natural soil itself.Gallego et al. (2008), quoted by Silva (2011), states that unpaved roads are characterised by low vehicle traffic, with a daily average of no more than 500 vehicles, and that their greatest demand is linked to agricultural activities, livestock farming and access to forest areas. Maintaining good road conditions is important for guaranteeing the country's economic and social progress. As such, Silva (2011) states that from a social point of view, unpaved roads are essential infrastructure for rural communities, due to the benefits they provide, since these roads offer people access to education, work and leisure, and are often the only way for rural communities to find essential services that are only available in urban centres.The physical condition of unpaved roads, according to Van Meerveld (2014), depends on factors specific to each region, such as soil type, climate, topography, traffic and maintenance frequency.According to Koetz (2009), one of the problems on unpaved roads is poor drainage, related to the poor sizing of drainage systems such as channels and drains, or the use of hydrological or soil data that is not consistent with the reality of the location.The erosion process due to surface runoff is one of the major problems associated with unpaved roads, causing soil loss and compromising roads as direct consequences. These roads are generally poorly designed and lack maintenance. Minella (2009) points to a number of factors responsible for the occurrence of erosion and sediment production on unpaved roads, including the intensity and duration of rainfall, characteristics of the material that makes up the road surface, road characteristics related to drainage, slope, traffic, road construction and maintenance, and external areas that drain onto the roads.An unpaved road in good condition must be strong enough to withstand traffic loads. To achieve this, it is necessary to adopt practices that, according to Minella (2009), should be preventative and, when necessary, maintenance, in order to reduce soil loss rates.

2.3 Conservation practices

Measures to control surface runoff and improve infiltration, combined with soil management with vegetation, are soil conservation practices that can reduce the problems caused by erosion on unpaved roads.

The choice of strategy depends on factors such as climate, topography, geology, soil type, as well as traffic intensity and the types of vehicles that travel the road most frequently (Cao et al., 2015). To this end, studies have been carried out to develop erosion models for unpaved roads, which estimate the amount of sediment produced as a result of the erosion process, contributing to the choice of appropriate management strategies. However, according to Fu et al. (2010) one of the limitations is access to quality data, since most modelling studies are aimed at assessing water quality aspects.In this context, due to the various controlling factors, the difficulty and the high costs of observing and measuring continuous erosion processes during runoff or erosion events due to the spatial and temporal scales at which the processes occur, effective models are needed to predict soil erosion and sediment input in different environmental contexts (Jaafari et al., 2015).

Cao et al. (2013) simulated sediment production and soil losses under five different rainfall intensities (43, 68, 83, 128, and 142 mm h^{-1}), three slope gradients (10-5%, 17-6% and 26-8%) for the development of a new model on unpaved roads on the Loess Plateau in China. In this study, runoff power can be a parameter for estimating soil losses in road surface gullies and based on other soil loss models, the authors developed a new equation according to rainfall intensity, runoff discharge and slope, finding that for this condition, an adequate drainage system and a reduction in the slope of the road can lead to a minimisation of sediment input.

Jaafari et al. (2015) developed an empirical model addressing sediment production and input form on dirt forest roads, as one of the indicators for assessing environmental impacts. The road erosion model WARSEM (Washington State Department of Natural Resources) was used to model sediment input rates and erosion estimation, and the main advantages achieved with the chosen model was the ability to deal with different environmental configurations, and simulate the effects of Best Management Practices (BMPs) of the related road, and can be used as a guideline for the development of long-term management plans.

Fu, Newham and Field (2009) applied the empirical WARSEM erosion model to unpaved roads, developed by the Washington State Department of Natural Resources, which is used to estimate the long-term average sediment input from roads to watercourses and considers data on road length, precipitation, the geology of the region, surface, slope, road width and traffic and the distance from the road and the stream. This data was modified according to the availability of data found for the south-eastern region of Australia and its application may be suitable for other regions. In addition, using this model it is possible to simulate different management conditions to reduce sediment input from roads into water bodies (Fu et al., 2009).

The development of a new infiltration basin model that includes the effects of spaced storms on the size of the basin was used to design a infiltration basin of a stretch of unpaved road in João Pinheiro, in the state of Minas Gerais, Brazil.

The model was evaluated in relation to infiltration rate, storm return period, and basin size parameters. The results showed that the infiltration rate was the most important variable affecting basin size and that small but widely spaced storm events influenced basin size more strongly than large single events (Silva et al., 2014).
Among the strategies to reduce sediment production and erosion on unpaved roads, the adoption of best management practices led to a reduction in sediment production of up to 80% on roads in the Oklahoma region, even with a limited budget (Turton et al., 2009). Effective Best Management Practices (BMPs) provided a stable surface with geotextile fabric and gravel, an adequate drainage bar for the road, stable channels to remove water without damaging the roadbed or cutting into the slope. This work suggests that where roads are deeply incised, a suitable and stable drainage bar cannot be provided, as the surface may become less effective with increasing storm intensity. Reductions in road sediment production and runoff can be especially effective in improving rural stream water quality, as road sediment is added directly to the region's streams, where impacts are greatest (Turton et al., 2009).
Avoiding the accumulation of water from surface run-off, whether generated on the road itself or on its banks, is a strategy for reducing erosion problems on unpaved roads by collecting this water and directing it to natural or artificial drains, accumulation basins or other retention systems located on the edge of the terrain (Griebeler et al., 2005).
In this context, a gully spacing model was developed by Griebeler et al. (2005) based on the characteristics of the areas that contribute to runoff (shear stress, soil resistance and channel slope) and through simulations, the model showed sensitivity to changes in characteristics such as erodibility and critical shear stress of the soil in addition to slope, and allowed changes in the tolerable deepening of the channel so that spacings can be widened or reduced.

3. FINAL CONSIDERATIONS

The importance of unpaved roads is essential for Brazil's socio-economic development, as they are the main route for transporting Brazilian agricultural production, as well as connecting rural populations to the nearest cities.
It can be seen that a large part of the road network is made up of unpaved roads, as they are quicker to build and cost less to implement than paved roads.
The maintenance and upkeep of these roads is of fundamental importance to society, as neglecting them causes a series of environmental damages, among

the most serious being the loss of soil and the formation of erosion processes. The main way in which erosion occurs in these areas is mainly due to poor drainage of water in the beds and banks of rural roads. Therefore, in order for the drainage system to be carried out properly, it is necessary to know the erodibility, water infiltration capacity in the soil and the adoption of mechanical practices in the beds and interception of rainwater through catchment devices.
In view of this, it was realised that a large number of rural roads do not have energy dissipaters, which together with the slope contributes even more to surface runoff, and that there is no management and handling of conservation practices, which if they were to be considered a priority in the design and maintenance of roads, would drastically reduce the problems faced with sediment production and soil loss, and consequently erosion.

REFERENCES

Cao L, Zhang K, Dai H, Liang Y. Modelling Interrill Erosion on Unpaved Roads in the Loess Plateau of China. L Degrad Dev. 2015;26:825-832.

Carvalho DF de, Cruz ES da, Pinto MF, Silva LDB, Guerra JGM. Rainfall characteristics and erosion losses under different soil management practices. Rev Bras Eng Agrícola e Ambient. 2009;13:3-9.

Casarin RD, Oliveira EL de. Erosion control on unpaved rural roads using a gradient terracing system associated with catchment basins. Irriga. 2009;14:548-563.

DNIT. National Road System [Internet]. Brasília, DF: National Department of Transport Infrastructure 2014 [accessed 13 July 2016]. Available at: http://www.dnit.gov.br/sistema-nacional-de-viacao/sistema-nacional-de-viacao

Enriquez AG, Silva DP, Pruski FF, Griebeler NP, Cecon PR. Erodibility and critical shear stress in the drainage channel of an unpaved rural road. Rev Bras Eng Agrícola e Ambient. 2015;19:160-165.

Fu B, Newham LTH, Field JB. Modelling erosion and sediment delivery from unsealed roads in southeast Australia. Math Comput Simul. 2009;79:2679-2688.

Fu B, Newham LTH, Ramos-Scharrón CE. A review of surface erosion and sediment delivery models for unsealed roads. Environ Model Softw. 2010;25:1-14.

Griebeler NP, Pruski FF, Da Silva JMA, Ramos MM, Da Silva DD. Model for determining the spacing between drains on unpaved roads. Rev Bras Cienc do

Solo. 2005;29:397-405.
Jaafari A, Najafi A, Rezaeian J, Sattarian A. Modelling erosion and sediment delivery from unpaved roads in the north mountainous forest of Iran. GEM - Int J Geomathematics. Springer Berlin Heidelberg; 2015;6:343-356.
Koetz M, Pruski FF, Mehl HU, Silva DD da, Marques EAG. Methods for determining soil erodibility and critical shear stress on unpaved roads. Reveng. 2009;110-119.
Liu YJ, Wang TW, Cai CF, Li ZX, Cheng DB. Effects of vegetation on runoff generation, sediment yield and soil shear strength on road-side slopes under a simulation rainfall test in the three gorges reservoir area, China. Sci Total Environ. Elsevier B.V.; 2014;485-486:93-102.
Minella JPG, Merten GH, Walling DE, Reichert JM. Changing sediment yield as an indicator of improved soil management practices in southern Brazil. Catena. Elsevier B.V.; 2009;79:228-236.
Morgan RPC. Soil erosion and conservation. 3° ed. Oxford: Blackwell; 2005.
Oliveira JF De, Griebeler NP, Correchel V, Silva VC Da. Erodibility and critical shear stress in unpaved road soils. Rev Bras Eng Agrícola e Ambient. 2009;13:955-960.

Oliveira JF De, Griebeler NP, Garvil R a, Oliveira JDM, Rabelo MWDO. Use of the Estradas software to determine the spacing between drains on unpaved roads in the interior of Goiás. Rev Ceres. 2011;58:17-22.

Rotta CM dos S. Study of the Recovery of Areas Degraded by Erosive Processes: Procedures and Efficiency of Methods. São Carlos: University of São Paulo; 2012.
Silva DP da. Model for dimensioning surface drainage systems on unpaved roads [thesis]. Viçosa: Federal University of Viçosa; 2011.
Silva DP da, Pruski FF, Agouridis CT, Márcio J, Denver W, Meireles M. Model for Designing Infiltration Basins in Tropical and Subtropical Climates with a Focus on Unpaved Roads. J Hidrol Eng. 2014;19:1-5.
Turton DJ, Smolen MD, Stebler E. Effectiveness of BMPS in reducing sediment from unpaved roads in the Stillwater Creek, Oklahoma Watershed. J Am Water Resour Assoc. 2009;45:1343-1351.
van Meerveld HJ, Baird EJ, Floyd WC. Controls on sediment production from an unpaved resource road in a Pacific maritime watershed. Water Resour Res. 2014;50:4803-4820.

CHAPTER 2

WATER REQUIREMENTS AND GROWTH RATES OF IRRIGATED SESAME CROP (Sesamum indicum L.) BRS ANAHÍ

Isaac Alves da Silva FREITAS Agronomist, PhD, UFERSA, Mossoró/RN. Ana Beatriz Alves de ARAÚJO Agricultural and Environmental Engineer, PhD, UFERSA, Mossoró/RN.

1 - INTRODUCTION

The edaphoclimatic conditions of the Brazilian semi-arid region are favourable to the sesame crop, as it is drought tolerant; it is an alternative of great economic and social importance for the semi-arid conditions of north-eastern Brazil, as it is easy to grow and generates income and work, as well as being a source of food for small and medium-sized producers.

Considering these aspects, it is essential to make technological adjustments to the sesame production process so that Northeastern producers can enjoy the potential benefits of this crop. In order to obtain higher yields and greater efficiency in the use of water, it is necessary to know the crop's water needs so that the plants can be offered the right amount of water for their metabolic and physiological processes (Azevedo et al., 2003).

Plant growth analysis is a method that describes the morphophysiological conditions of plants at each time interval within their cycle. Knowing the growth characteristics and phenology of a cultivated plant is important for carrying out actions such as planning crop treatments and evaluating performance, as well as for making decisions (MARUR and RUANO, 2001).

The aim of this study was to analyse the main growth and development parameters of BRS Anahí sesame, relating them to its water requirements.

2. THEORETICAL BASIS

2.1 Species description

Sesame (Sesamum indicum L.) is a dicotyledonous plant of the Pedalaceae family and is one of the oldest oil plants used by humankind. Sesame is composed of a stem that varies in height from 0.5 to 3 metres and can be erect, with or without branching and with or without hairs, expressing a level of heterophily. The sesame stem, usually 15 cm from the soil surface, has different shapes, square and rounded, on which the leaves grow, showing a reasonable level of heterophily: those at the bottom are larger, wider and rounder, and those at the top are narrower and longer, to allow the sun's rays to penetrate all the leaves, while the intermediate leaves do not shade the lower ones (ANDRADE, 2009).On both sides it is covered in a thick cuticle with hairs. It has a pivoting root system, with the presence of a main root, an axonorphic system, whose branches are poorly developed, penetrating the soil perpendicularly. Sesame has two flowers per leaf axil, they are complete, gamopetalous, zygomorphic and appear in clusters, alternating or opposite. Each flower has a short peduncle in the axils (YERMANOS, 1980). Sesame fruits are hairy capsules, varying in size from 2 to 8 cm in length and in diameter from 0.5 to 2 cm. They are located in the leaf axils, and in each leaf axil there can be from one to three fruits (varietal characteristic), and the number of fruits per plant is directly related to the number of flowers. The dehiscence of the fruit begins at the apex, towards the base. The seeds are small and 1,000 seeds weigh between 2 and 4g, with more than 20 per fruit locule. They are oval, slightly flattened and vary in colour from white to black, brown, olive green and yellow. There are differences between cultivars in terms of the number of seeds per fruit and the size of the seeds. Their main constituent is oil which, depending on the cultivar and location, can exceed 60 per cent of the seed's weight (BELTRÃO, 2001).

2.2 Growth analysis

Growth analysis produces knowledge of practical value and accurate information (CARDOSO et al., 1987) regarding the growth and behaviour of genotypes, which can be used for the benefit of growers, enabling them to choose the cultivar that best suits each region. Growth analysis is a method that follows the dynamics of photosynthetic production and is vitally important for understanding the plant's morphophysiological processes and their influence on yield. It can also be used to determine the net production of plants, derived from the photosynthetic process, as a result of the performance of the assimilatory system over a given period of time (CARDOSO et al., 1987); it also allows the physiological processes of plant growth and development to be analysed. The main indicators of plant yield capacity and efficiency are: Plant Height, Leaf Diameter, Number of Leaves, Leaf Area (LA), Leaf Area Index (LAI); Crop Growth Rate (CGR); Relative Growth Rate (RGR); Net Assimilation Rate (NIR) (BENINCASA, 2003), among others.

2.3 Water requirements

Among the various uses of water resources, irrigation stands out for its socio-economic importance in arid and semi-arid agricultural regions, where it is practised to supplement natural rainfall in meeting the water needs of crops (FARIAS et al., 2000). According to Bernardo (2005), it is also necessary to know the behaviour of the crop as a function of the different amounts of water supplied and to identify the stages of development with the highest water consumption and the critical periods when a lack or excess would cause production to fall. Although sesame is a drought-tolerant plant, its highest yields will be seen when grown under favourable water conditions (SOUZA et al., 2000).According to Beltrão et al. (2001) there is little information on the water requirements of sesame, mainly because it is rarely grown on a commercial scale under irrigation. The lack of information is not limited to only the total amount of water required, but also its consumption in the different phonological phases.

2.4 Evapotranspiration (ET) and Reference Evapotranspiration (ETo)

The term reference evapotranspiration (ETo), using the concepts proposed by Thornthwaite and applying them to the alfalfa crop (Mendicago sativa L.), with a height of 0.3 to 0.5 metres, in a given climate condition, with a buffer area of approximately 100 metres. Among the reasons given for choosing alfalfa as the reference plant, it was argued that its height would be more representative of the aerodynamic conditions of other crops than grass with its low height. However, Doorenbos and Pruitt (1977) presented a definition for reference evapotranspiration as that which occurs on an extensive grassy surface, with a height of 0.08 to 0.15 m, in active growth, completely covering the soil surface and without water restrictions. In essence, it can be seen that the definition of reference evapotranspiration coincides with potential evapotranspiration (ETP).
The crop coefficient (Kc) is an empirical relationship between the evapotranspiration of a crop (ETc), under conditions of no water stress, and the reference evapotranspiration (ETo). This coefficient reports the phenological and physiological development of a particular crop in relation to the reference evapotranspiration and also represents the water use of a specific crop, which is of relevant importance for estimating its water requirement, necessary for both the sizing of the irrigation system and the operationalisation of irrigated

perimeters (CLARK et al. 1996; MOHAN; ARUMUGAM, 1994).Among researchers in various countries around the world, the Penman-Monteith method for obtaining ETo, described by Allen et al. (1998), has been accepted as the standard method for calculating Reference Evapotranspiration.Using the Penman-Monteith method, it is possible to calculate both the hourly evapotranspiration and then arrive at the daily evapotranspiration by adding up the hourly values, as well as calculating the daily evapotranspiration using the average daily data.

3. MATERIAL AND METHODS

The experiment was conducted at the experimental station belonging to the Agricultural Research Company of Rio Grande do Norte EMPARN, in the municipality of Apodi, RN, whose geographical coordinates are: latitude 5° 37' 37" S, longitude 37° 49' 54" W and altitude 138m, whose predominant climate is warm BSW'h' according to the Koppen classification. The crop used in the experiment was BRS Anahí sesame, planted on 6 August 2015 with a spacing of 0.8 m between rows and 0.1 m between plants, which corresponds to a linear planting density of 10 plants/m. Sprinkler irrigation was carried out to meet the crop's water demand, three times a week, on Monday, Wednesday and Friday.

Every 15 days, 20 plants were randomly sampled in the field and some growth parameters were analysed non-destructively. In the laboratory, some analyses were carried out using methodologies considered destructive. The parameters analysed were:Average plant height (cm) - This measurement was taken considering the distance between the soil surface and the upper end of the main stem of the plants;

Stem diameter (cm) - measured at 4 cm from the ground using a digital metal caliper accurate to 0.1 mm;

Leaf area - Estimated using the dry mass of leaf discs method, based on the ratio of the dry mass of the leaf discs (9 discs) to the total dry mass of the leaves. The leaf discs were obtained using a perforator of known area.

The crop's water requirements were estimated based on the reference evapotranspiration (ETo) which, in turn, was calculated using meteorological data collected at the National Meteorological Institute (INMET) weather station installed at EMPARN's research base. The method used was the Penman-Monteith/FAO model, defined by the equation:

$$ET_0 = \frac{0{,}408\Delta(R_n - G) + \gamma\left(\frac{900\,U_2}{T+273}\right)(e_s - e_a)}{\Delta + \gamma(1 + 0{,}34U_2)}$$

Where: Rn is the radiation balance, G is the ground heat flux, both expressed in MJm^{-2} day^{-1} , γ is the psychrometric constant, Δ is the declination of the water vapour saturation curve (KPa $°C^{-1}$) and U_2 is the wind speed (daily average) ($m.s^{-1}$) at 2m above the ground surface; es: Daily average vapour saturation pressure (kPa); and_a : Daily average current vapour pressure (kPa); T: Daily average air temperature at 2m height (°C).

The crop coefficients (Kc) used for each development stage were obtained from PEREIRA et al. (2014). The crop evapotranspiration (ETc) at each phenological stage was obtained from the Kc and ETo data, according to the ratio:

ETc = ETo x Kc

Where: ETo is the reference evapotranspiration in mm. Day^{-1} , obtained using the FAO Penman-Monteith method; Kc crop coefficients at each phenological stage.

4. RESULTS AND DISCUSSION

The vegetative cycle of BRS Anahí sesame was divided into four phases according to the occurrence of the main physiological processes, as shown in Table 1.

Table 1. Phenological cycle of sesame, cultivar BRS Anahí in the various stages of development, in the municipality of Apodi-RN, 2015.

	Stages of development	Days after emergence (DAE)	Duration (Days)
Phase I	Emergence at 10% ground cover	1 - 20	20
Phase II	10% ground cover at the start of flowering	21 - 44	24
Phase III	Start of flowering to start of ripening	45 - 68	24
Phase IV	Beginning to end of ripening	69 - 82	14
	TOTAL		82

The data shown in Table 1 is data observed during the period of the experiment and characterised by observing points that are characteristic of each stage of the crop.Looking at Figure 1, it can be seen that there was slow growth at the start of vegetative development, followed by marked growth in the middle phase and a tendency to stabilise in the final phase. The equation representing the relationship between average plant height and days after sesame emergence can be seen in Figure 1, showing a $y = -0.0006x^3 + 0.0841x^2 - 1.2021x + 6.4$ $R^2 = 0.9909$ and an R= 0.99543, which indicates a good correlation between the parameters.analysed. It can be seen that between 30 and 60 days after emergence, the height of the plant increases sharply, and after this period, the growth of the crop slowed down compared to the start of development, remaining close to 125 cm. Grilo and Azevedo (2013) when assessing the sesame cultivar BRS Seda in Ceará-Mirim-RN, observed a maximum average height of 170.0 cm. This difference may be due to changes in the architecture of the two varieties.

Figure 1. Average height of sesame plants in Apodi (RN), 2015.

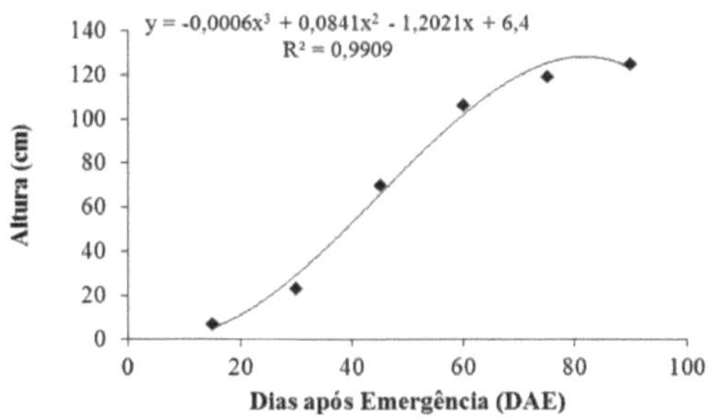

The average stem diameter values, in mm, are shown in Figure 2. The average stem diameter was 9.69 mm. It was observed that there was a marked increase in plant diameter up to the 45th day of the crop cycle. The statistical coefficient of determination of $R^2 = 0.9983$ confirms an excellent correlation between these two parameters analysed.

Figure 2. Average stem diameter (mm) of sesame, in Apodi (RN), 2015.

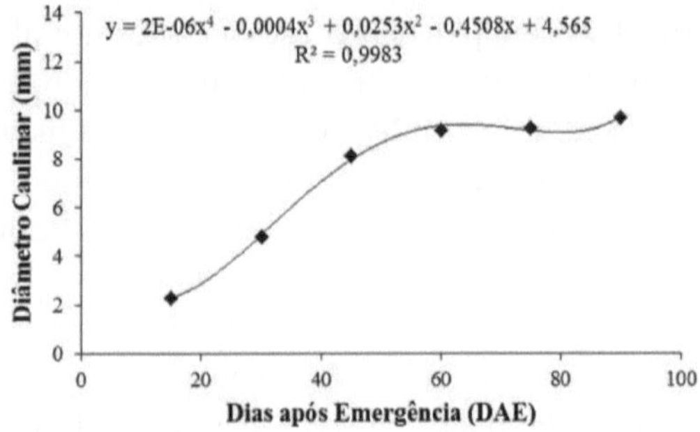

The evolution of the crop's leaf area throughout its vegetative cycle can be seen in Figure 3. The curve that best fitted the leaf area (Af) as a function of the days after emergence (DAE) was that of a 4th degree polynomial. The crop grew rapidly between the 35th day after emergence and the 75th day and decreased after that. The leaves during the aforementioned days, because they are large and broad, contribute more weight to the leaf area, while at the end of the cycle, heterophily gives the plant narrower leaves, which causes this leaf area decay behaviour. The correlation coefficient reached 1, showing an excellent relationship between the two parameters studied.

Figure 3 Leaf area per sesame plant throughout the phenological cycle in Apodi (RN), 2015.

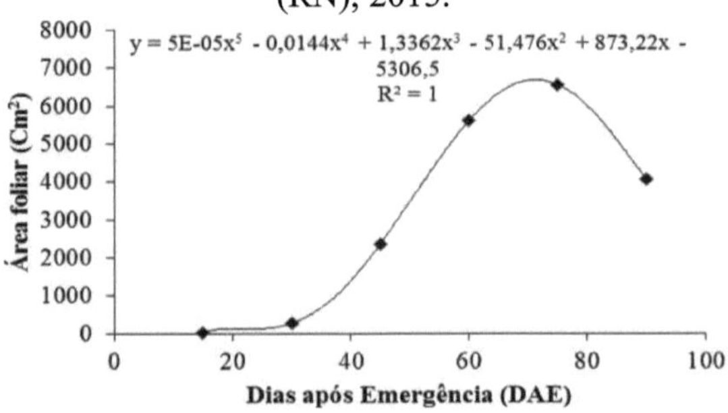

Figure 4 shows the daily variations in ETc, as well as the ETo values estimated by the FAO Penman-Monteith method. There was a continuous increase in the evapotranspiration of the sesame crop in relation to ETo throughout its vegetative cycle. ETo remained more or less constant throughout the experimental period, with an average daily value of around 7.8 mm. The crop's evapotranspiration behaviour was increasing, but quite variable throughout its phenological cycle, with a minimum value of 3.5 mm d^{-1} and a maximum of 8.4

mm d^{-1} . The low ETc values at the beginning and end of the cycle are related to the low leaf area index in the initial phase and the crop's senescence in the final phase. According to Souza (2003), variations in the daily evapotranspiration rate during the same phenological phase can be due to local meteorological conditions and the duration of that phase. Thus the result found by Souza in 2003 corroborates the behaviour of ETc found in this study, both in terms of the difference between ETc in the different phenological phases and the behaviour of Etc within the phase itself, which have approximate values between the days in each phase.

Figure 4 - Variation in sesame crop evapotranspiration (ETc) and reference evapotranspiration (ETo) in Apodi (RN), 2015.

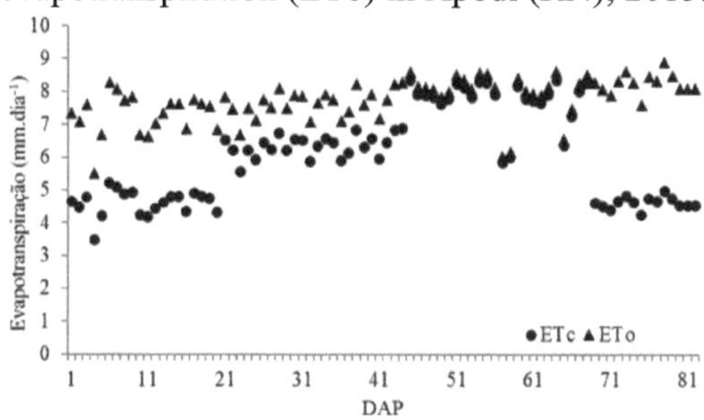

When analysing the average daily values of crop evapotranspiration for each phase of the phenological cycle during the observation period, there was a minimum value of 4.6 mm d^{-1} at the start of vegetative development (Phase I) and a maximum of 7.7 mm d^{-1} in the flowering and fruit formation phase (Phase III), with an average value for the entire cycle of 6.0 mm d^{-1} . It can also be seen that the crop's evapotranspiration increased from Phase I to Phase III, when it reached its maximum, and then decreased again. The sesame's total water consumption throughout the growing cycle was 494.0 mm.

5. CONCLUSIONS

The development cycle of BRS Anahí was completed early when compared to the crop average of 90 days. With regard to the growth and development of the crop, there was a reasonable statistical correlation between the parameters plant height, stem diameter and leaf area, allowing these parameters to be estimated as a function of the number of days after plant emergence under regional conditions.The crop's water needs were represented by its evapotranspiration (ETc), which showed low values at the beginning and end of the growing season, indicating low water consumption. The greatest demands were made during the flowering and fruit formation phase, when ETc was 8.4 mm.day $.^{-1}$ For the crop's vegetative cycle, water consumption was 494 mm of irrigation.

REFERENCES

ANDRADE, P.B. Potential pollinators and pollination requirements of sesame. 2009. 74f. Dissertation (Master's in Animal Science) - Centre for Agricultural Sciences, UFC, Fortaleza, 2009.

AZEVEDO, P. V. DE; SILVA, B. B. DA; SILVA, V. DE P. R. Water requirements of
irrigated mango orchards in northeast Brazil. Agricultural Water Management, v.58, p.241-254, 2003.

BELTRÃO, N.E.M.; SILVA, L.C.; QUEIROGA, V.P.; VIEIRA, D.J. Soil preparation,
fertilisation and liming. In: BELTRÃO, N.E.M; VIERA, D.J. (Org.). The sesame agribusiness in Brazil. EMBRAPA, Brasília. Informação Tecnológica, 2001. p.109-131.

BLANC, E.; QUIRION, P.; STROBL, E. The climatic determinants of cotton yields: evidence from a plot in West Africa. Agricultural and Forest Meteorology, v.148, p.1093-1100, 2008.

BENINCASA, M. M. P. Analysing plant growth: basics. Jaboticabal: FUNEP, 2003. 41 and 42 p.

BERNARDO, S. Irrigation manual. 7 ed. current. expanded. Viçosa: UFV, 2005. 611 p.

CARDOSO, M. J.; FONTES, L. A. N.; LOPES, N. F.; et al. Assimilate partitioning and dry matter production of maize in two association systems with beans (Phaseolus vulgares L.). Revista Ceres, Viçosa, v. 34, n. 191, p. 71-89, 1987.

CLARK, ALBREGTS, E. E; STANLEY, C. D. et al. Water requirements and crop coefficients of drip-irrigated strawberry plants. Transaction of ASAE, St. Joseph, v. 39,n. 3, p. 905-912, 1996.

DOORENBOS, J.; PRUITT, J. O. Guidlines for predicting crop water requirements. Rome, IT: FAO, 1977. 179 p. (FAO. Irrigation and Drainage Paper, 24).

FARIA, R. A. de; SOARES, A. A .; SEDIYAMA, G. C.; RIBEIRO, C. A. A. S. Demand for supplementary irrigation for maize crops in the state of Minas Gerais. Brazilian Journal of Agricultural and Environmental Engineering, v.4, n.1, p.46-50, 2000.

MARUR, C. J.; RUANO, O. A. A reference system for determination of developmental states of uplant cotton. Revista de Oleaginosas e Fibrosas, Campina Grande, v.5, n.2, p. 313-317, 2001.

PEREIRA, R.P.; BELTRÃO, N. E.M; ARRIEL, H.C. A.; SILVA, E.S.B. Fertilisation
organic sesame in the Seridó Paraibano. Revista de Oleaginosas e Fibrosas. Revista de Oleaginosas e Fibrosas, v.6, n.3, p.599-608, 2002.

LIMA, V.I. Growth and production of sesame cv. G3 as a function of zinc and boron. 2006. 72p. Dissertation (Master's degree in agronomy) - Federal University of Paraíba. Areia, PB, 2006,72p.

MESQUITA, J.B.R. Management of the sesame crop subjected to different irrigation rates, doses of nitrogen and potassium by the conventional method and by fertigation. 2010. 82f. Dissertation (Master's degree in agronomy) - Federal University of Ceará, Fortaleza, 2010.

MOHAN, S.; ARUMUGAM, N. Crop coefficient so major crops in south India. Agricultural Water Management, Amsterdam, v. 26, p. 67-80, 1994.

SOUZA, C. B. de. Water requirements, growth and development of the pineapple tree in the tabuleiros of Santa Rita - PB. Campina Grande: UFPB, 2003. 124 p. Doctoral thesis.

SOUZA, J. G. de; BELTRÃO, N. E. M.; SANTOS, J. W. dos S. Physiology and sesame yields in soil with water deficiency. Revista Brasileira de Oleaginosas e Fibrosas, Campina Grande, v. 4, n. 3, p. 163-169, 2000.

YERMANOS, D.M. Sesame. In: FEHR, W.R.; HADLEY, H.H. Hybridisation of crop plants. Madson, Wisc., ASA. P.549-563. 1980.

CHAPTER 3
ORGANIC WASTE ADDUCTION: PHYSIOLOGICAL AND BIOCHEMICAL RESPONSES IN SUNFLOWER CROPS (HELIANTHUS ANNUUS L.)

1 - INTRODUCTION

Maria da Conceição Freitas MOURA Agronomist, PhD, UFERSA, Mossoró/RN. Alek Sandro DUTRA Professor, PhD, UFC, Fortaleza/CE. Rosilene Oliveira MESQUITA Professor, PhD, UFC, Fortaleza/CE.

Sunflower (Helianthus annuus L.) stands out for its desirable agronomic characteristics, such as its short cycle and wide adaptability to different soil and climate conditions (SANTOS et al., 2015). It also enhances subsequent crops by improving the chemical, physical and biological conditions of the soil (FERRARI and SOUZA, 2009).Sunflower species have high quality oil for biofuel production and human consumption, as well as for animal feed, due to the generation of by-products from the oil extraction process. It also stands out for being a grain and fodder producing species (LIRA et al., 2011). It also stands out as an option for producers in the crop rotation system. It affects soil fertility, providing a high capacity for nutrient cycling and a low rate of nutrient export (FEITOSA et al., 2013). According to Santos et al. (2012) the sunflower crop has agronomic characteristics such as tolerance to cold, drought and qualitative characteristics, which makes it a favourable crop to plant in the Northeast region.

Several studies have evaluated the improvement of the soil's chemical properties through the use of organic sources in agriculture. One of the alternative organic sources that deserves to be analysed in order to observe its efficiency on crops would be the use of the residue generated from the processing of roasted coffee beans, which, through chemical analyses, has been found to have high mineral levels, especially of potassium, which is an essential nutrient for plant development. The use of organic fertilisers can make nutrients available t o plants and microorganisms (KRAY et al., 2011), improve the chemical and physical attributes of the soil (SAMPAIO et al., 2012) by adding organic matter, and provide greater recycling of nutrients in the soil by crops compared to the use of mineral fertilisers (MELO et al., 2011). Organic alternatives for nutrients in agriculture not only minimise the environmental impact of improperly disposing of waste in the soil, but also reduce production costs, equalling or surpassing the grain yield of crops obtained through the use of chemical fertilisers (STROJAKI et al., 2013). In addition to the sustainability promoted in agricultural environments, organic sources enable a better form of exploitation and quality of what is being generated or produced on agronomic soils, because by using organic sources, they pollute the growing environment less, enabling quality as a whole. Sunflowers are demanding when it comes to potassium, so the availability of potassium in the soil for sunflower production should be medium to high, as its demand is high. According to Sfredo et al. (1984), sunflower extracts 40% more K than soya and maize crops. According to Hafsi et al. (2014), potassium is a macronutrient that participates in many physiological processes, such as osmotic adjustment, photosynthesis, transport and enzyme activation in plants. Potassium is important for the movement of solutes via the phloem, acting indirectly to increase the rate of photosynthesis (KARLEY and WHITE, 2009).The aim of this study was to assess the physiological and biochemical characteristics of two sunflower cultivars that were fertilised with organic waste with a high potassium content.

2 - MATERIAL AND METHODS

The experiment was conducted in the experimental area of the Agriculture Sector of the Phytotechnics Department of the Agricultural Sciences Centre of the Federal University of Ceará (UFC), located on the Pici Campus in Fortaleza, Ceará, between April and August 2015. The chemical characteristics of the soil taken from the 0-20 and 20-40 cm layers before the experiment was set up are shown in Table 1. The following were sampled 20 random points in the area, mixing them and obtaining a composite soil sample.

Table 1. Chemical and granulometric analysis of the soil collected in the 0-20 and 20-40 cm layers. Fortaleza, Ceará. 2015.

pH P K Na Mg Al Ca M.OSilt Clay

Layer (cm)		(mg/kg)		-------------- (cmolc/kg)- -----------			(g/kg)	-------- (g/kg)--- ------	
0 - 20	(2015)	6,8	6	0,17 0,13	0,70 0,15	1,00	7,45	416 78	64
20- 40		6,7	6	0,17 0,20	0,60 0,20	0,60	5,79	50881	90

Source: Soil and Plant Analysis Laboratory of the Soil and Plant Nutrition Department of the Federal University of Ceará, Fortaleza - CE.

The residue used in the treatments, whose chemical characteristics are shown in Table 2, was obtained after the coffee bean processing and roasting process at a temperature of 270°C, generating an increasing volume of ash, which was the material used in the analyses.

Table 2. Chemical analysis of the organic waste used in the treatments.

Chemical characteristics of organic waste

.... (g/kg)..................................							.... (mg/kg) ....
N	M.O	P	P2O5	K	K2O	CaMg	Fe C Zn Mn
12,9	331,2	7,1	16,3	83,3	101,6	48,6 28,7	4632, 9216,9 71,9 219,5

Source: Soil and Plant Analysis Laboratory of the Soil and Plant Nutrition Department of the Federal University of Ceará, Fortaleza - CE.

The waste had a high moisture content, so before using it in the treatments, it was laid out on benches to dry for a period of one week in a greenhouse at a temperature ranging from 28 to 30 °C. After this period, it was stored in sealed plastic bags to avoid contamination and then used as fertiliser to grow the plants. The soil in the experimental area was prepared in the conventional way, consisting of ploughing and two cross harrows; the seeds were sown using of sunflower cultivars BRS 122 and BRS 323, with fertilisation based on soil analysis and following the recommendations of Lira et al. (2011), using urea as the source of nitrogen (60 kg ha^{-1} of N) divided into two doses, the first in foundation at the time of planting and the second 35 days after planting in top dressing, along with boric acid (2 kg ha^{-1} of B). The source of phosphorus was simple superphosphate at a dose of 80 kg ha^{-1} of P2O5 applied once at the time of planting; the crop was planted in single rows with a spacing of 0.7 m between rows x 0.3 m between plants. The experimental area consisted of 40 plots with a size of 11.2 m^2 , each experimental plot consisted o f 4 rows totalling 48 plants, with the centre rows being considered the useful area (evaluation areas). Both treatments had four replications corresponding to four blocks.The experimental design used was randomised blocks in a 2x4+2 factorial scheme, with two cultivars, four doses of organic residue (T1=0, T2=40, T3=80 and T4=120 kg ha^{-1} of K) plus the additional control, which was potassium chloride at a dose of 60 kg ha^{-1} of K for both cultivars.

Leaf temperature and rates of photosynthesis (A), transpiration (E), stomatal conductance (gs), internal CO2 concentration (Ci) and the ratio between internal and external CO2 concentration (Ci/Ca ratio) were measured on two fully expanded leaves from the middle third of the plants in the useful area at 45 days after sowing, using LICOR's Li - 6400XT (Portable Photosynthesis System - LI) infrared gas analyser (IRGA). At 50 days after sowing, two leaves were collected from the middle third of the plants located in the useful area and then lyophilised and macerated to obtain the extract that was used to determine N-aminosolubles, protein concentration and soluble carbohydrates.

The determination of N-aminosolubles was carried out according to the method of Yemm and Cocking (1955). 500μL of the leaf extract, 250 μL of citrate buffer (0.2M, pH 5.0), 500 μL of KCN (0.2 Mm) and 100 μL of 5% ninhydrin were added to test tubes, then the tubes were closed, shaken and placed in a water bath for 20 minutes at 100° C, after which the reaction was stopped and the tubes were placed in an ice bath. After cooling, 650 μL of 60% ethanol was added. The levels of aminosoluble N were estimated using readings of absorbance at 570 nm based on a standard curve adjusted from increasing concentrations of glycine. The soluble protein content was determined according to the Bradford methodology (1976). 1.0 mL of Coomassie's reagent was added to 100 □L of the diluted extract. This reagent was prepared by dissolving 100 mg of "Coomassie Brilliant Blue G-250" in 50 mL of 95% ethanol plus 100 mL of 85% phosphoric acid. The final volume of the solution was topped up to 1 L with deionised water. After mixing, the tubes were placed in a shaker for 1 hour, after which time the tubes were left to stand for 15 minutes. The soluble proteins were estimated by measuring the absorbance at 595 nm and expressed as mg g^{-1} MS. Bovine serum albumin was used as a standard. Soluble carbohydrates were determined according to Dubois et al. (1956). The reaction mixture consisted of 200 □L of the extract, 200 □L of 5% phenol and 1.0 mL of concentrated sulphuric acid. The mixture was then shaken vigorously and left to stand at room temperature (25° C) for 20 minutes to cool. The soluble carbohydrates were then quantified using absorbance measurements at 490 nm, using a curve made with increasing concentrations of anhydrous D(+) glucose as a standard. The data obtained was subjected to analysis of variance with Tukey's test at 5% probability to compare means, using ESTAT software (ESTAT, 1994).

3 - RESULTS AND DISCUSSION

The results showed that there were no significant differences between the organic residue and potassium chloride in terms of stomatal conductance (gs), net photosynthesis (A), transpiration rate (E), internal carbon concentration (Ci), internal and external carbon ratio (Ci/Ca ratio) and leaf temperature (TF). Stomatal conductance (gs) is shown in Figure 1, which compares the doses of organic residue with the control for the BRS 122 and BRS 323 cultivars, as a function of the mean and standard deviation. Thus, in the cultivars analysed (BRS 122 and BRS 323) at the dose of 80 kg ha^{-1} of K from the organic residue, they showed a conductance of 1.19 mmol m^{-2} s^{-1} and 1.25 mmol m^{-2} s^{-1} , respectively, while the values of 1.09 mmol m^{-2} s^{-1} for cultivar BRS 122 and 1.06 mmol m^{-2} s^{-1} for cultivar BRS 323. Evaluations carried out by Gondim et al. (2015) on the effects of manure doses in the presence and absence of mineral fertiliser on gas exchange in beetroot plants found superior stomatal conductance in the presence of mineral fertiliser, which was explained by the presence of potassium. Thus, in this study, it was found that stomatal conductance was not altered by the doses of organic residue and potassium chloride.Prazeres et al. (2015), when analysing the growth and gas exchange of cowpea plants under salinity and potassium doses, found that the stomatal conductance of cowpea plants increased with increasing potassium doses, a fact that was not observed in this research, since as the dose increased twice the recommended one (120 kg ha^{-1}), stomatal conductance decreased for both cultivars (Figure 1), but did not differ significantly between the other doses and the control. According to Catuchi et al. (2012), potassium fertilisation increased stomatal conductance in soybeans because K can provide plants with adequate stomatal conductance dynamics, as well as being an extremely important element in activating the carboxylase function of Rubisco, which contributes to greater photosynthetic activity (PRADO, 2008). According to Epstein and Bloom (2006), potassium is one of the most important nutrients for plants, as it participates in processes such as the opening and closing of stomata, photosynthesis, carbohydrate transport and respiration.

Figura 1 - Valores da condutância estomática (*gs*) de folhas de cultivares de girassol. (▭) testemunha (60 kg ha^{-1} de K); (▭) 0 kg ha^{-1} K (▭) 40 kg ha^{-1} de K do resíduo orgânico; (▭) 80 kg ha^{-1} de K do resíduo orgânico; (▭) 120 kg ha^{-1} de K do resíduo orgânico.

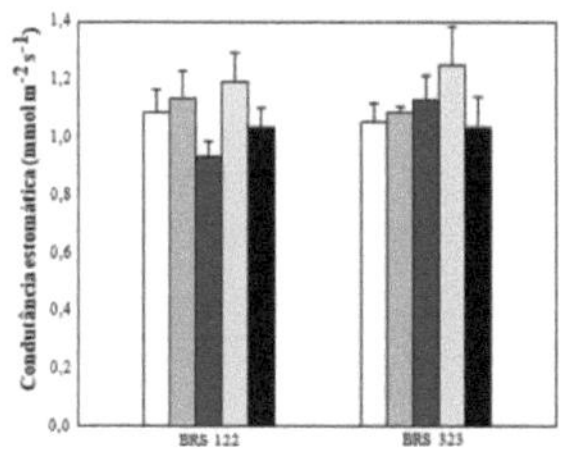

Some studies show that potassium deficiency affects plant gas exchange, especially stomatal conductance and Rubisco activity (JIA et al. 2008). According to Paiva et al. (2005), stomatal conductance regulates gas exchange and is therefore directly related to the photosynthetic process and consequent plant growth and development. The rate of net CO2 assimilation, transpiration, stomatal conductance and intercellular CO2 concentration are correlated parameters that can be used to diagnose physiological changes in plants when they are subjected to adverse conditions such as low or high amounts of nutrients (GONDIM et al., 2015). According to Taiz and Zeiger (2013), the inadequate supply of essential elements to plants causes disturbances in metabolic processes, which results in abnormal plant functioning.

In relation to the net photosynthesis rate (A), Table 4 compares the dose of 80 kg ha^{-1} of K with the control. These results show that the plants fertilised with the dose of 80 kg ha^{-1} of K from the organic residue, ranged from 41.31 μmol m^{-2} s^{-1} in the BRS 122 cultivar to 40.50 μmol m^{-2} s^{-1} for the BRS 323 cultivar, while the control showed a photosynthetic rate ranging from 38.70 μmol m^{-2} s^{-1} to 36.46 μmol m^{-2} s^{-1} in the BRS 122 and BRS 323 cultivars, respectively. According to Sultana et al. (1999), a reduction in stomatal conductance can reduce the photosynthetic rate by reducing the partial pressure of CO2 in the intercellular spaces. This information is also mentioned by Taiz and Zeiger (2013) when they state that greater stomatal opening favours the entry of CO2 into the leaf mesophyll, which increases the internal concentration and consequently photosynthesis.

Table 4: Net photosynthesis (A) of sunflower plants as a function of organic and mineral fertilisation.

Net photosynthesis (A) (µmol m^{-2} s^{-1})		
Doses (kg ha)$^{-1}$		Cultivars
BRS		122BRS 323
8041	.31 A40	.50 A
Test (60)	38.70 A36	.46 A

*Means followed by the same letter do not differ according to Tukey's test at 5% probability.

According to Amaral et al. (2006), monitoring stomatal dynamics makes it possible to understand physiological processes, since stomata are the main pathway for gas exchange between the atmosphere and the interior of the photosynthetic apparatus. Studies carried out by Jia et al. (2008) with rice plants found that the photosynthetic rate increased with an adequate supply of potassium compared to plants deficient in this nutrient, which was attributed to an increase in the carboxylation activity of Rubisco. The reduction in photosynthetic efficiency can be attributed to stomata not opening, as potassium plays an important role in maintaining the turgidity of guard cells, while potassium deficiency increases the intracellular concentration of CO2 (CAVALCANTE et al., 2015).Analysing the transpiration rate (E) between the cultivars fertilised with organic residue at a dose of 80 kg ha^{-1} of K compared to those fertilised with potassium chloride, it was found that the dose of 80 kg ha^{-1} of K from the organic residue for both cultivars (BRS 122 and BRS 323), respectively, showed transpiration of 19.41 to 14.86 mmol of H2O m s^{-2} $^{-1}$ (Table 5). As for the plants fertilised with potassium chloride, their behaviour varied from 14.55 mmol of H2O m s^{-2} $^{-1}$ for the BRS 122 cultivar and 14.40 mmol of H2Om s^{-2} $^{-1}$ for the BRS 323 cultivar.

Table 5: Transpiration rate of sunflower plants fertilised with organic waste and ineral fertiliser.

Transpiration rate (E) (mmol of H2O m^{-2} s)$^{-1}$

Doses (kg ha)-1		Cultivars	
	BRS 122		BRS 323
80 Test (60)	19,41 A 14,55 A		14,86 A 14,40 A

*Means followed by the same letter do not differ according to Tukey's test at 5% probability.

The reduction in transpiration rate, according to Pinto et al. (2008), is also directly linked to a reduction in stomatal conductance. According to Machado et al. (2005), even if there are fewer stomatal openings, there will be a reduction in transpiration, with a consequent increase in leaf temperature. This result was also observed in the present study, as the control plants showed lower values for stomata conductance, transpiration, photosynthesis and a growing increase in leaf temperature, when compared to the plants fertilised with organic waste in the dose of 80 kg ha^{-1} of K. Taiz and Zeiger (2013) further reinforce the above information, since according to these authors the increase in temperature can be caused by a reduction in transpiration, a consequence of the reduction in stomatal conductance, showing a smaller opening of the stomata, which restricts the dissipation of heat along with the loss of water vapour. Hussain et al. (2015) studied the influence of potassium fertilisation on growth, physiology and nutrient absorption in maize and found that a dose of 130 kg ha^{-1} of K promoted maximum net photosynthesis, transpiration and stomatal conductance.As gas exchange is regulated through stomatal movement, the absorption of external CO2 promotes greater transpiration, likewise, the reduction in transpiration limits the entry of carbon into the substomatal chamber (SHIMAZAKI et al., 2007), so it is necessary for plants to absorb the maximum amount of CO2 with the minimum loss of water (TAIZ and ZEIGER, 2013). Also according to Taiz and Zeiger (2013), significant variations in plant gas exchange are influenced by climatic conditions; however, nutritional supply is closely related to variations in gas exchange. As for the quantification of N-aminosolubles, there was no significant difference between the treatments and the control (plants fertilised with KCl). Figure 2A compares the different doses of organic residue with the control for both cultivars. N-aminosolubles (organic solutes) act in the plant by promoting osmotic adjustment as well as in processes related to stomatal opening, cell growth and elongation.The same pattern of behaviour was seen for the total soluble protein content (Figure 2B), which was statistically the same between the treatments and the control. This shows that all the potassium sources applied were effectively absorbed and used to produce amino acids and then proteins. The production of correctly structured proteins and their maintenance are important for efficient cellular function in the plant (LAWLOR, 2002).

Figura 2 - Concentração de N-aminosolúveis e proteínas em folhas de cultivares de girassol. ( ) testemunha (60 kg ha^{-1} de K); ( ) 0 kg ha^{-1} K; ( ) 40 kg ha^{-1} de K do resíduo orgânico; ( ) 80 kg ha^{-1} de K do resíduo orgânico; ( ) 120 kg ha $^{-1}$ de K do resíduo orgânico.

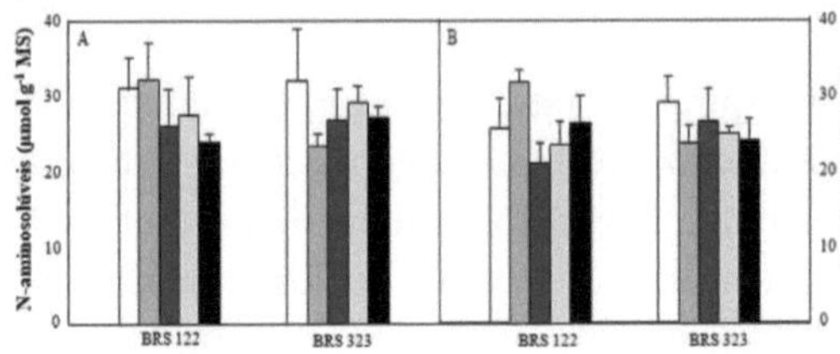

In relation to soluble carbohydrates, there was no significance between the treatments compared to the control (Figure 3). According to Prado (2008), potassium is an important nutrient for the formation of carbohydrates in the leaves and plays a fundamental role in translocating these assimilates to the various parts of the plant. In studies carried out by Nascimento et al. (2014), analysing the total soluble carbohydrates in leaves and roots of jatoba seedlings (Hymenaea courbaril L.) subjected to suppression of N, P and K, they observed that there were no significant differences between the treatments in relation to the analysis of the leaves, they only found a significant difference for the carbohydrate content in the roots, the authors were able to verify that with the absence of potassium there was a greater accumulation of carbohydrates in the root system.

Figura 3 - Concentração de carboidratos solúveis em folhas de cultivares de girassol. ( ) testemunha (60 kg ha^{-1} de K); ( ) 0 kg ha^{-1} de K; ( ) 40 kg ha^{-1} de K do resíduo orgânico; ( ) 80 kg ha^{-1} de K do resíduo orgânico; ( ) 120 kg ha^{-1} de K do resíduo orgânico.

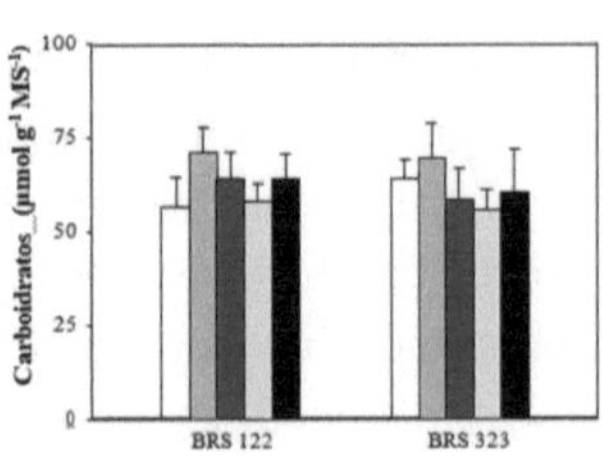

In plants deprived of potassium, carbohydrates begin to accumulate in the chloroplasts and this leads to a decline in the biochemical phase of photosynthesis (KARLEY and WHITE, 2009). According to Epstein and Bloom (2006), the accumulation of carbohydrates in the leaves and the decline in photosynthesis are indicative of potassium deficiency. In this study, the treatment with no potassium fertilisation had the highest carbohydrate values when compared to the control treatment and the other treatments (Figure 3).

4 - CONCLUSION

Fertilising sunflower plants with organic waste did not influence their physiology, and it performed as well as potassium chloride.

REFERENCES

AMARAL, J. A. T. do; RENA, A. B.; AMARAL, J. F. T do. Seasonal vegetative growth of coffee trees and its relationship with photoperiod, fruit set, stomatal resistance and photosynthesis. Pesquisa Agropecuária Brasileira, v.41, n.3, p.377-384, 2006.

BRADFORD, M.M. A rapid and sensitive method for the quantification of microgram quantities of protein utilizing the principles of protein-dye binding. Analytical Biochemistry, v.72, n.1/2, p.248-254, 1976.

CATUCHI, T.A.; GUIDORIZZI, F.V.C.; GUIDORIZI, K.A.; BARBOSA, A.M.;SOUZA, G.M. Physiological responses of soya cultivars to potassium fertilisation under different water regimes. Pesquisa Agropecuária Brasileira, v.47, n.4, p.519-527, 2012.

CAVALCANTE, V. S.; PRADO, R. M.; ALMEIDA, H. J.; CRUZ, F. J. R.; SANTOS,D. M. M. Gaseous exchanges, growth and foliar anatomy of sugarcane plants grown in potassium (K) deprived nutrient solution. Australian Journal of Crop Science, v.9, n.7, p.577-584, 2015.

DUBOIS, M., K. A. GILLES, J. K. HAMILTON, P. A. REBERS, F. SMITH. Colourimetric method for determination of sugars and related substances. Analytical Biochemistry, v.28, n.3, p.350-356, 1956.

EPSTEIN, E.; BLOOM, A. J. Nutrição mineral de plantas: princípios e perspectivas. 2° ed. Londrina: Planta, 2006. 392p.

ESTAT. System for statistical analyses (v. 2.0). Jaboticabal: Department of Exact Sciences, FCAV-UNESP, 1994.

FEITOSA, H. O.; FARIAS, G. C.; SILVA JÚNIOR, R. J. C.; FERREIRA, F. J.; ANDRADE FILHO, F. L.; LACERDA, A. F. Influence of borax and potassium fertilisation on sunflower performance. Comunicata Scientiae, v.4, n.3, p.302-307, 2013.

FERRARI, R. A.; SOUZA, W. L. Evaluation of the oxidative stability of sunflower oil biodiesel with antioxidants. Revista Química Nova, v. 32, n. 1, p. 106-111, 2009.

GONDIM, A. R. O.; SANTOS, J. L. G.; LIRA, R. P. L.; BRITO, M. E. B.; PEREIRA,F. H. Photosynthetic activity of beetroot subjected to mineral fertilisation and cattle manure. Green Journal of Agroecology and Sustainable Development, v.10, n.2, p.61- 65, 2015.

HAFSI, C.; DEBEZ, A.; ABDELLY, C. Potassium deficiency in plants: effects and signalling cascades. Acta Physiologiae Plantarum, v.36, n.5, p.1055-1070, 2014.

HUSSAIN, A.; ARSHAD, M.; AHMAD, Z.; AHMAD, H. T.; AFZAL, M.; AHMAD, M. Potassium fertilisation influences growth, physiology and nutrients uptake of maize (Zea mays L.). Cercetări Agronomice în Moldova, v.68, n.1, p.37-50, 2015.

JIA, Y.; YANG, X.; ISLAM, E.; FENG, Y. Effects of potassium deficiency on chloroplast ultrastructure and chlorophyll fluorescence in inefficient and efficient genotypes of rice. Journal of Plant Nutrition, v.31, n.12, p.2105-2118, 2008.

KARLEY, A. J.; WHITE, P.J. Moving cationic minerals to edible tissues: potassium, magnesium, calcium. Current Opinion in Plant Biology, v.12, n.3, p. 291-298, 2009.

KRAY, C. H.; TEDESCO, M. J.; BISSANI, C. A.; BORTOLON, L.; ANDREAZZA, R.; GIANELLO, C. Evaluation of the application of urban waste compost and sewage sludge on two different soils. Pesquisa Agropecuária Gaúcha, v.17, n.2, p.119-125, 2011.

LAWLOR, D. W. Carbon and nitrogen assimilation in relation to yield: mechanisms are the key to understanding production systems. Journal of Experimental Botany, v.53, n.370, p.773- 787, 2002.

LIRA, M. A.; CARVALHO, H. W. L.; CHAGAS, M. C. M.; BRISTOT, G.; DANTAS,J. A.; LIMA, J. M. P. Avaliação das potencialidades da cultura do girassol, como alternativa de cultivo no semiárido nordestino, EMPARN: Natal-RN, 2011. 40 p.

MACHADO, E. C.; SCHMIDT, P. T.; MEDINA, C. L.; RIBEIRO, R. V.

Photosynthesis responses of three citrus species to environmental factors. Pesquisa Agropecuária Brasileira, v.40, n.12, p.1161-1170, 2005.

MELO, A. V.; GALVÃO, J. C. C.; BRAUN, H.; SANTOS, M. M.; COIMBRA, R. R.; SILVA, R. R.; REIS, W. F. Nutrient extraction and biomass production of black oats grown in soil subjected to eighteen years of organic and mineral fertilisation. Semina: Ciências Agrárias, v.32, n.2, p.411-420, 2011.

NASCIMENTO, H. H. C.; PACHECO, C. M.; LIMA, D. R. M.; SILVA, E. C.; NOGUEIRA, R. J. M. C. Ecophysiological aspects of Hymenaea courbaril L. seedlings in response to N, P and K suppression. Scientia Forestalis, v.42, n.103, p.315-328, 2014.

PAIVA, A. S.; FERNANDES, E. J.; RODRIGUES, T. J. D.; TURCO, J. E. P. Stomatal conductance in bean leaves subjected to different irrigation regimes. Engenharia Agrícola, v.25, n.1, p.161-169, 2005.

PINTO, C. de M.; TAVORA, F. J. F. A.; BEZERRA, M. A.; CORRÊA, M. C. de M. Growth, distribution of the root system in peanut, sesame and castor bean to cycles of water deficiency. Revista Ciência Agronômica, v.39, n.3, p.429-436, 2008.

PRADO, R. M. Nutrição de plantas. Ed. UNESP. São Paulo, SP, 2008. 407p.

PRAZERES, S. da S.; LACERDA, C. F.; BARBOSA, F. E. L.; AMORIM, A. V.;ARAÚJO, I. C. da S.; CAVALCANTE, L. F. Growth and gas exchange of cowpea plants under saline irrigation and potassium doses. Revista Agro@mbiente on-line, v.9, n.2, p.111-118, 2015.

SAMPAIO, T. F. et al. Sewage sludge in the recovery of degraded areas: effect on soil physical characteristics. Revista Brasileira de Ciência do Solo, v. 36, n. 5, p. 1637- 1645, 2012.

SANTOS, E. R.; BARROS, H. B.; CAPONE, A.; FERRAZ, E. C.; FIDELIS, R. R.Effect of sowing times on sunflower cultivars in the south of the state of Tocantins. Revista Ciência Agronômica, v.43, n. 1, p. 199-206, 2012.

SULTANA, N.; IKEDA, T.; ITOH, R. Effect of salinity on photosynthesis and dry matter accumulation in developing rice grains. Environmental and Experimental Botany, v. 42, n. 3, p. 211-20, 1999.

SFREDO, G.J., CAMPO, R.J.,SARRUGE, .R.Sunflower: mineral nutrition and fertilisation. (Technical circular, 8) Embrapa-CNPSo, Londrina, Brazil. 36p. 1984.

SHIMAZAKI, K. I.; DOI, M.; ASSMANN, S. M.; KINOSHITA, T. Light regulation of stomatal movement. Annual Review of Plant Biology, v.58, n.6, p.219- 247, 2007.

STROJAKI, T. V.; SILVA, V. R.; SOMAVILLA, A.; DA ROS, C. O.;

MORAIS,
M.T. Soil chemical attributes and sunflower and maize productivity as a function of the application of urban waste compost. Pesquisa Agropecuária Tropical, v. 43, n.3, p.278-285, 2013.
TAIZ, L.; ZEIGER, E. Plant physiology, 5.ed. Porto Alegre: Artmed, 2013. 954 p.YEMM, E. W.; COCKING, E. C. The determination of aminoacids with ninhydrin. Analyst, v.80, n.948, p.209-214, 1955.

CHAPTER 4

CHEMICAL CHANGES IN SOIL IRRIGATED WITH DILUTIONS OF PRODUCED WATER TREATED IN A GREENHOUSE

Ricardo André Rodrigues FILHO Agricultural and Environmental Engineer, Master's Degree, UFERSA, Mossoró/RN.
Ana Beatriz Alves de ARAÚJO Agricultural and Environmental Engineer, PhD, UFERSA, Mossoró/RN.

1. INTRODUCTION

The importance of water is decisive for biodiversity on the planet, as it conserves and balances the life of the most diverse species. In the historical context, it was the search for water that led man to occupy territories, develop agriculture and improve his quality of life. However, changes in man's lifestyle and his relationship with nature have meant that over the years, water has come to be seen as a water resource and no longer as a source of life.

Population growth, together with the advance of industrial practices, has led to an increase in water scarcity in regions that did not have this problem. Allied to this, anthropogenic factors have over the years caused a series of environmental impacts linked to water and soil contamination. As a result, various studies are being carried out to mitigate these environmental impacts.

Work related to the use of oil produced water for irrigation shows that this can be a viable alternative for the destination of this wastewater after its proper treatment (MIRANDA et al., 2016; CRISÓSTOMO et al., 2018).

Miranda et al. (2016) found that castor beans irrigated with produced water treated with reverse osmosis did not affect the salinity or sodicity of the soil, while water treated by filtration showed negative results in terms of crop yield and an increase in soil salinity and sodicity. Crisóstomo et al. (2018) irrigated ornamental pineapple trees with filtered produced water and found that there was a considerable increase in the risk of soil salinisation and sodification. However water treated with reverse osmosis applied to the soil in a short cycle did not affect the physico-chemical characteristics of the soil. The aim of this study was to assess the effects of treated produced water diluted with well water on the chemical attributes of a typical dystrophic red argysoil cultivated with sunflower.

2. LITERATURE REVIEW

2.1 CHARACTERISATION OF OIL PRODUCED WATER

Oil wastewater is defined as the effluent generated from oil and natural gas production processes. It receives this name when it is found on the surface of the ground together with the material that has been extracted from the reservoir and consists of a mixture of formation water from the oil-producing well, condensation and injection water from secondary recovery processes and water used to desalinise the oil produced (FAKHRU'L-RAZIA et al., 2009).

For offshore exploration, the wastewater is treated and disposed of at sea. In onshore production, the wastewater can be reinjected into the wells, which can lead to a decrease in productivity (MOTTA et al., 2013). This practice may not be economically viable, given its high cost and the fact that it should only be carried out when the underground structure is capable of accommodating the injected water.

According to Curbelo (2002), a large amount of the water produced cannot be reused in other production processes, so the water that has not been reused is sent for treatment in order to adapt it to environmental standards for disposal in rivers or the ocean. Some of the water can be reused within the production fields, serving to reinject it into the reservoirs to keep the pressure high so that more oil can be produced (BADER, 2007).

2.2 COMPOSITION OF PRODUCED WATER

The constitution of water is a complex mixture of naturally occurring organic and inorganic chemical compounds that have been dissolved or dispersed in the form of particulates from geological formations and migration routes where this effluent has been dammed for thousands of years (NEFF et al., 2011).

Compared to seawater, produced water has the same salts and heavy metals, but in much higher concentrations. This can be explained by the formation time of the geological structure in which the material is found (GABARDO, 2007). The most common metals found are barium (Ba) and iron (Fe), chloride ions (Cl^-), sodium (Na^+), calcium (Ca^+), magnesium (Mg^{2+}), ammonia and sulphide are the most abundant inorganic compounds in produced water (OGP, 2005).

Produced water also contains a wide variety of heavy metals, some in high concentration and some suspended elements such as sand, silt, silicates, clays and traces of radionuclides (WANG et al., 2012).

2.3 CHANGES IN THE QUALITY OF SOIL IRRIGATED WITH PRODUCED WATER

Some studies carried out using treated produced water to irrigate soil cultivated with different crops have shown satisfactory results in terms of soil quality at the end of the cycle. Miranda et al. (2016) irrigated castor beans (Ricinus communis) with produced water and concluded that water treated with reverse osmosis had no effect on the productivity of the castor bean crop, did not affect the salinity and sodicity of the soil and concluded that irrigation can be used without restriction. However, irrigation with water treated by filtration showed negative results, reducing productivity by 30 per cent compared to irrigation with well water for papaya and increasing salinity and sodicity levels in a Quartzarenic Neosol.

Crisóstomo et al. (2018) carried out work at Fazenda Belém, Aracati - CE, belonging to the Potiguar basin. Produced water was used to irrigate soil cultivated with ornamental pineapple (Ananas comosus var. bracteatus) and they found that irrigation with filtered produced water considerably increased the risk of soil salinisation and sodification. On the other hand, soil irrigated with reverse osmosis treated produced water, in a short cycle, proved to be a viable alternative for ornamental pineapple production.

Pinheiro et al. (2014), in another study carried out at Fazenda Belém, assessed the effects of produced water on the irrigation of soil cultivated with oilseeds and flowers ornamental plants. As a result, they found that the water produced had to undergo a series of treatments, such as filtration and desalination, to make it viable for use in agricultural activity. Water quality is based on the basic criteria of salinity, sodicity and toxicity.

3. MATERIAL AND METHODS

3.1 LOCATION AND CHARACTERISATION OF THE EXPERIMENTAL AREA

The work was carried out in a vegetation house located in the Department of Agronomic and Forestry Sciences at the Centre for Agrarian Sciences of the Federal Rural University of the Semi-Arid Region - UFERSA, Mossoró/RN (05 12' 03.9" LS; -37 19' 37.98" LO). Figure 1 shows the location of the greenhouse where the experiment was carried out. According to Alvares et al. (2013), the region's climate is hot and dry, classified according to Köppen-Geiger as semi-arid BSh. It has an average annual temperature of over 26.5 ºC and average annual rainfall of 794 mm, which is irregularly distributed and concentrated between three and six months of the year, and relative humidity of 68.9%.

Figure 1 - Location of the experimental area at UFERSA, Mossoró campus/RN.

Source: Google maps (2019).

3.2 Collection of treated oil produced water (AT) and well water (AP)

Produced water was collected three times from reservoirs with a capacity of 1,000 litres. A company located in the region that operates in the area of oil production provided the water for the experiment. The collected produced water was treated at the site where the experiment was set up, using an organic, strongly cationic polymer called AGEFLOC DW-3753. It acts in aqueous solution and serves to flocculate suspended material, has a high coagulation power and accelerates the speed of floc formation. In its treatment, 40ml of the polymer was applied for every 1,000 litres of water produced.

3.3 Characterisation of the soil used in the experiment

The soil used to fill the pots was taken from the Rafael Fernandes Experimental Farm belonging to UFERSA and located in the district of Alagoinha, Mossoró, RN. According to the Brazilian Soil Classification System (EMBRAPA, 2013), the soil collected for the experiment is classified as Typic Dystrophic Red Argissolo, as classified by Rêgo et al (2016). Table 1 and 2 show the physical and chemical attributes of the soil before the dilutions of produced water were applied. Soil samples were taken at depths of 0 to 0.10m and 0.10 to 0.20m.

Table 1. Chemical characteristics of the soil before irrigation.

Profundidade (m)	pH	CE_{es} dSm^{-1}	MO $g\ kg^{-1}$	N $g\ kg^{-1}$	P	K^+	Na^+	Ca^{2+}	Mg^{2+}	Fe	Mn	Zn	Cu
					.......mg dm^{-3}.......			$cmol_c dm^{-3}$		...........mg dm^{-3}.............			
0 a 0,10	6,1	0,41	3,31	0,35	1,9	26,9	4,7	1,1	0,7	1,56	0,56	0,2	0,01
0,10 a 0,20	4,8	0,03	2,07	0,21	1,7	26,9	4,7	0,4	1,0	1,95	0,55	0,2	0,02

Profundidade (m)	Al^{3+}	H+Al	SB	T	CTC	V	m	PST
		..........cmolc dm^{-3}..........				.........%.........		
0,00 a 0,10	0,0	1,49	1,89	1,89	3,37	56	0	1,00
0,10 a 0,20	0,15	1,97	1,49	1,64	3,47	43	9	1,00

Note: pH - Hydrogenionic Potential; ECes - Electrical Conductivity in the Saturation Extract; OM - Organic Matter; N - Nitrogen; P - Phosphorus; K - Potassium; Na - Sodium; Ca - Calcium; Mg - Magnesium; Fe - Iron; Mn - Manganese; Zn - Zinc; Cu - Copper; Al - Aluminium; H+Al - Potential acidity; SB - Base sum; t - Effective cation exchange capacity; CEC - Cation exchange capacity; V - Base saturation; m - Aluminium saturation; PST - Percentage of exchangeable sodium. **Source:** Researcher's collection (2019).

Table 2. Soil physical characteristics before irrigation.

Profundidade (m)	Densidade do solo	Massa Especifica de partículas	Areia	Silte	Argila	Classe Textural	Relação Silte/Argila
	$g\ cm^{-3}$	$g\ cm^{-3}$		.... kg kg^{-1}..........			
0,00 a 0,10	2,78	1,78	0,9	0,03	0,07	Areia	0,44
0,10 a 0,20	2,70	1,72	0,84	0,03	0,13	Areia Franca	0,24

Source: Researcher's collection (2019)

3.4 Conducting the experiment

The experiment was conducted using a randomised block design with twenty-five plots, five treatments and five replications, as shown in Figure 1.

Figure 2 Schematic of the experimental area.

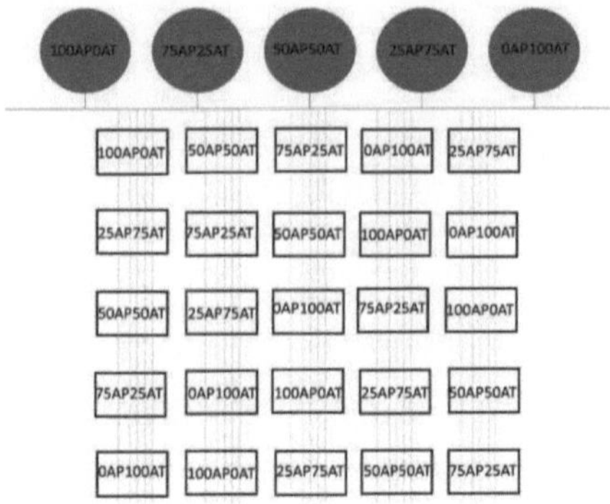

Source: Researcher's collection (2019).

The treatments were defined according to the different dilutions of produced water treated with well water. Table 3 shows the treatments with their respective dilutions.

Table 3. Dilutions of treated produced water (AT) in well water (AP) used in the experiment.

Dilutions	Dilutions
100AP0AT	100% well water and 0% treated produced water
75AP25AT	75% well water and 25% treated produced water
50AP50AT	50% well water and 50% treated produced water
25AP75AT	25% well water and 75% treated produced water
0AP100AT	0% well water and 100% treated produced water

Source: Author (2019).

3.5 Setting up the experimental plots and transplanting the seedlings

The sunflower crop used in the experiment was the cultivar H-360 (Heliantus annus), from Embrapa Semiárido. Transplanting took place 12 days after sowing, where the seedlings were taken to the field after sowing in polythene trays. To ensure successful transplanting and survival of the seedlings, irrigation with dilutions of treated produced water was only started 20 days after cultivation; previously, irrigation was only by water supply. The spacing used in the experiment was 1.0 m between rows of plants and 0.60 m between plants. The 25 pots received the seedlings on 14 September 2017. The pots were filled with zero gravel and sieved Typic Dystrophic Red Argisol collected from the UFERSA Experimental Farm. The plots were filled with foundation fertiliser in all the pots with a single application of urea, MAP, potassium chloride and boric acid.The sunflower's water demand was determined using adjustment coefficients based on the reference evapotranspiration (ETo). To determine the gross water table, a water balance was carried out, where the water input was irrigation and the output was crop evapotranspiration (ETc).All the meteorological data needed to determine the need for irrigation was collected from a weather station installed in the greenhouse where the experiment took place. The sunflower cycle was 90 days after transplanting (DAT), and a total of 435.35 mm of gross water was applied for each day. treatment, where the amount of produced water and supply water was divided according to the percentage for each dilution.

3.6 Chemical attributes of the soil after applying dilutions of treated produced water

The soil was chemically characterised by collecting samples from each pot at depths of 0 to 0.10m and 0.10 to 0.20m, following the recommendations proposed by Teixeira et al. (2017), where the following analyses were carried out: hydrogen potential (pH) using a pH meter, electrical conductivity of the saturation extract (ECes) using a conductivimeter, sodium (Na^+) using a flame photometer, calcium (Ca^{2+}) and magnesium (Mg^{2+}) using the titrimetric method.

3.7 Statistical analysis

To assess the soil data before irrigation with the dilutions, a randomised block design was used with five replicates and five treatments, where the plots were the treatments and the sub-plots were the depths at which the samples were taken, from 0 to 0.10m and 0.10 to 0.20m. With regard to the soil characteristics, the data was subjected to analysis of variance using the SISVAR (Systems for Analysis of Variance) computer programme developed by Ferreira (2011), using

Tukey's test and orthogonal contrasts for multiple comparisons of the means. Costa et al. (2018) proposed a combination of the contrasts with the treatment averages, expressed as follows: 1) treatment T5 versus the other treatments (T1, T2, T3 and T4); 2) treatment 4 versus treatments T1, T2 and T3; 3) treatment T3 versus treatments T1 and T2 and 4) treatment T2 versus treatment T1. Thus, the contrasts used were:

$C1 = 12\ m1 + 12\ m2 + 8\ m3 + 10\ m4 - 42\ m5$

$C2 = 30\ m1 + 30\ m2 + 20\ m3 - 80\ m4$ $C3 = 6\ m1 + 6\ m2 - 12\ m3$

$C4 = 6\ m1 - 6\ m2$

4. RESULTS AND DISCUSSION

4.2 CHEMICAL CHARACTERISATION OF SOIL IRRIGATED WITH DILUTIONS OF PRODUCED WATER TREATED WITH WELL WATER

Table 4 shows the statistical analysis of the Tukey test carried out on the soils that received irrigation with the treated produced water treatments.

Table 4. Chemical characteristics of soil irrigated with dilutions of produced water treated with well water

Diluições	Características químicas do solo				
	pH	CE_{es}	Na^+	Ca^{2+}	Mg^{2+}
		dS m^{-1}	mg dm^{-3}	......$cmol_c$ dm^{-3}......	
100AP0AT	7,49	0,47	259,56	1,01	0,58ª
74AP25AT	7,54	0,50	307,44	1,08	0,74ab
50AP50AT	7,73	0,51	279,17	1,19	0,85ab
25AP75AT	7,70	0,45	332,95	1,25	0,95ab
0AP100AT	7,57	0,59	310,83	1,12	1,05b
Média	7,60	0,50	297,99	1,13	0,838
CV 1 (%)	4,35	29,46	50,49	21,94	32,81
CV 2 (%)	5,54	19,12	42,35	25,0	34,47
Erro-padrão	0,104	0,047	47,57	0,078	0,087
Probabilidade	NS	NS	NS	NS	0,013*
Probabilidade dos contrastes (decimal)					
Contraste C_1: 0AP100AT Vs (100AP0AT + 75AP25AT + 50AP50AT + 25AP75AT)	NS	NS	NS	NS	0,009**
Contraste C_2: 25AP75AT Vs (100AP0AT + 75AP25AT + 50AP50AT)	NS	NS	NS	NS	0,002**
Contraste C_3: 50AP50AT Vs (100AP0AT + 100AP0AT)	NS	NS	NS	NS	NS
Contraste C_4: 75AP25AT Vs 100AP0AT	NS	NS	NS	NS	NS

Note: pH - Hydrogen potential; ECes - Electrical conductivity of the soil saturation extract; K^+ - Potassium; Na^+ - Sodium; Ca^{2+} - Calcium; Mg^{2+} - Magnesium. C1 = 12 m1 + 12 m2 + 8 m3 +10 m4 - 42 m5; C2 = 30 m1 + 30 m2 + 20 m3 - 80 m4; C3 = 6 m1 + 6 m2 - 12 m3; C4 = 6 m1 - 6 m2

Averages followed by equal letters in the column do not differ statistically by the Tukey test at 5% probability. and** * Significant at 1% and 5% probability by the F test, respectively.

NS - Not significant at 5% probability by F test. **Source:** Researcher's collection (2019).

Analysing Table 4, it can be seen that the attributes pH, electrical conductivity in the saturation extract, calcium and sodium did not show a significant effect at 5% probability using the Tukey test. The result of the statistical analysis showed that only for magnesium (Mg^{2+}) there was a significant effect at 5% probability using the Tukey test. The treatment containing only well water (100AP0AT) had a lower pH value than the others because the water had a lower concentration of salts than the other treatments. Compared to the initial state of the soil before the irrigation took place, it can be seen that the pH in the soil increased after irrigating it with treated produced water because the condition of the dilutions already had a high pH. Soil pH is divided according to agronomic classes for interpreting soil acidity. Values below 4.5 are considered very low; between 4.5 and 5.4 low; 5.5 to 6.0 good; 6.1 to 7.0 high and above 7.0 is considered very high. Considering the specifications for agricultural and forestry enterprises, the pH found in the treatments is considered very high (RIBEIRO et al., 1999).The treatments showed no statistical variation for the orthogonal contrasts in relation to ECes. Overall, an average of 0.50 dS m^{-1} was obtained. Statistically, this attribute had no significant effect using the Tukey test.According to Miranda et al. (2016), the sunflower crop is considered moderately sensitive to salinity. The crop's productivity is associated with the salt content in the soil and irrigation water, where the EC must not exceed 1.1 dS m^{-1} so as not to reduce its yield. Na concentrations^{+} in the soil ranged from 259.56 mg dm^{-3} (100AP0AT) to 332.95 mg dm^{-3} (25AP75AT). The average sodium content in the treatments was 297.99 mg.dm^{-3} . There was no statistical variation between the orthogonal contrasts, nor in the treatments. High levels of Na^{+} in the soil can cause structural problems for the soil, affecting pores and helping to alter infiltration and hydraulic conductivity (VELOSO, 1991). The dispersion of clay in the soil can cause a reduction in the stability of soil aggregates, where the clay fraction can be easily transported to more superficial areas of the soil (SILVA, 2003). The average Ca^{2+} values in the soil ranged from 1.01 $cmol_c$ dm^{-3} for the 100AP0AT to 1.25 $cmol_c$ .dm^{-3} for the 25AP75AT. The soil fertility interpretation class proposed by Ribeiro et al. (1999) shows that the concentrations of Ca^{2+} in the soil irrigated with the different dilutions of treated produced water are in the range considered low in relation to exchangeable calcium in the soil, with only the 25AP75AT being classified as medium fertility. It was found that the dilutions of treated produced water in supply water had a significant effect at the 5% probability level using the Tukey test on the Mg attribute^{2+} in the soil. The results showed an increase in exchangeable magnesium in the soil for each increase in the amount of treated produced water placed in the dilution, with the 100AP0AT treatment obtaining the lowest value with 0.58 $cmol_c$ dm^{-3} and T5 the highest concentration with 1.05 $cmol_c$ dm .$^{-3}$ The C1 and C2 contrasts were significant at the 1% probability level, where the 0AP100AT treatment differed from the sum of the means of the other treatments and the 25AP75AT treatment had a different Mg concentration^{2+} from the sum of the 100AP0AT, 75AP25AT and 50AP50AT treatments. Following the soil fertility interpretation class for the cation exchange complex of Ribeiro et al. (1999), magnesium is classified as very low (< 0.15 $cmol_c$ dm^{-3}), low (0.15 to 0.45 $cmol_c$ dm^{-3}), medium (0.46 to 0.90 $cmol_c$ dm^{-3}), good (0.91 to 1.50 $cmol_c$ dm^{-3}) and very good (>1.50 $cmol_c$

dm^{-3}). The average concentrations of exchangeable magnesium found in the soils irrigated with the different dilutions are as follows: treatment 100AP0AT is considered very low; 75AP25AT and 50AP50AT are of medium fertility, 25AP75AT and 0AP100AT are of good fertility.
Based on the analysis of the soil's chemical characteristics after irrigation with dilutions of treated produced water over a 90-day cycle, it can be said that soil quality assessments and monitoring should be carried out constantly, given that the higher the irrigation rate applied to the soil and the crop cycle, the greater the concentration of the attributes analysed.

4.3 Analysis of Hydrogen Potential (pH) and Electrical Conductivity of the Soil Saturation Extract (CEes), Calcium (Ca^{2+}), Magnesium (Mg^{2+}) and Sodium (Na^{+}) in the soil profile at the end of the sunflower cycle.

Figure 2 shows the pH, ECes, Ca^{2+} , Mg^{2+} and Na^{+} values at the depths of 0 to 0.10m and 0.10 to 0.20m. Figure 8A shows that the soils with the lowest pH values were those irrigated only with well water and all the treatments fall within the agronomic classification of Ribeiro et al. (1999) as very high pH. With regard to pH behaviour in the soil profile, there was a tendency for its value to decrease in deeper layers. Miranda et al. (2016) in their work irrigating sunflower crops with treated produced water in three cycles found that soil pH varies according to the crop cycle and the type of produced water treatment used. The treatment using reverse osmosis was found to have lower values than the water collected from the Açu aquifer and the produced water treated using the filtration method.
In general, there was a downward trend in the average pH values, probably due to the leaching of salts into the subsurface layers, as well as a variation in the clay content of each soil layer, resulting in the possible sorption of salts (COSTA, 2017).

Figure 2: pH values (A), ECes (B), Ca^{2+} (C), Mg^{2+} (D) and Na^{+} (E) of the soil under produced water application as a function of depth for each treatment.

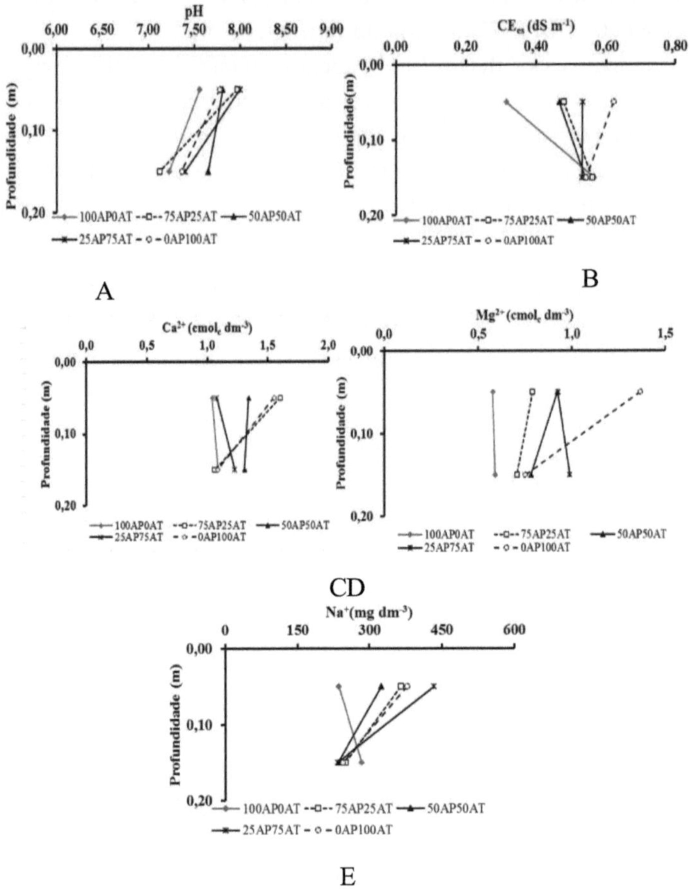

Source: Researcher's collection (2019).

Soil salinity has a number of impacts on plant development, which can progressively increase the concentration of salts in the roots and cause yield problems for crops that are sensitive to high salt concentrations. In Figure 2B, only the 100AP0AT treatment showed leaching characteristics due to the irrigation process draining the salt levels from the soil surface. The other treatments showed little variability in electrical conductivity as a function of depth. Ayers and Westcot (1999) mention that in the topsoil some factors can cause salts to accumulate, such as: evaporation of water on the soil surface, leading to a higher concentration of salts; continuous leaching and replacement of salts in the vicinity of the wet zone.The 100AP0AT, 75AP25AT, 50AP50AT and 25AP75AT treatments showed a small increase in electrical conductivity along the soil profile. The 0AP100AT treatment, on the other hand, behaved differently, with a higher salt concentration in the surface layers. The result found in this study corroborates Costa (2017), where the application of dilutions of dairy wastewater increased the electrical conductivity of the soil in layers with higher clay contents and due to leaching.

This tendency for the EC to increase with an increase in the amount of treated produced water disposed of in the soil is due to the increase in the value of the electrical conductivity in the dilution waters, thus showing results similar to those of Miranda et al. (2016), where irrigation with produced water treated by filtration increased the EC_{es} in relation to the soil irrigated only with supply water practically throughout its profile. Figure 2C shows the behaviour of Ca^{2+} concentrations in the soil profile. The treatment containing only well water showed lower concentrations of Ca^{2+} throughout the soil profile. Most treatments (75AP25AT, 0AP100AT and 50AP50AT) showed higher calcium concentrations in the surface layers, where 75AP25AT and 0AP100AT showed a considerable decrease as the depth increased. A similar result was found by Crisóstomo et al. (2018), where the concentrations of exchangeable calcium and magnesium were higher in the top layer of soil irrigated with produced water treated by osmosis.For exchangeable magnesium, only the 0AP100AT treatment reduced considerably along the profile (Figure 8D); the other treatments did not show considerable fluctuations. The increase in Mg^{2+} in the soil was proportional to the application of treated produced water in the dilutions. Considering that the 0AP100AT treatment has the highest concentration of Mg^{2+} in the soil, most of this element moved to the surface zone because the organic matter retained this nutrient. The results found in this study corroborate the values of exchangeable magnesium in the soil found by Crisóstomo et al. (2018), where the concentrations of Mg^{2+} were higher in the surface layers. Figure 2E shows that the average concentrations of Na^{+} in most of the treatments showed a reduction in their values as a function of depth, with only the 100AP0AT treatment showing a tendency for the concentration to increase in the deeper layers. According to NSW (2010), in high concentrations, sodium can cause problems in soil structure, such as dispersion, reduced water infiltration, forming a compacted layer and intensifying erosion processes. The evaporation of water in the soil caused the accumulation of Na^{+} in the surface layers in most treatments. Crisóstomo et al. (2018), found that in the irrigated ornamental pineapple cycle, produced water treated by reverse osmosis provided a lower concentration of Na^+ in the soil than produced water treated by filtration. NSW (2010) explains that soils that have had prolonged irrigation with wastewater can present problems in the future, as inadequate soil management combined with a lack of agronomic recommendations have adverse effects on soil structure, causing clay dispersion, drainage problems and damaging crop development.

5. CONCLUSIONS

The analyses carried out on the soil after irrigation with the dilutions of treated produced water in supply water showed that of all the chemical attributes assessed, only Mg^{2+} had a significant effect. Its concentration in the soil increased along with the increase in treated produced water applied. Analysis by orthogonal contrasts revealed that there was an increase in magnesium in the 25AP75AT and 0AP100AT treatments.The 75AP25AT dilution was the most recommended for irrigation purposes, as it minimised the chemical changes in the Argissolo studied. Finally, it should be emphasised that other variables such as heavy metals and hydrocarbons present in the produced water should be monitored to prove that applying the water to the ground is an efficient method of disposing of the effluent generated in oil production.

REFERENCES

AYERS, R. S.; WESTCOT, D. W. Water quality in agriculture. 2.ed. Campina Grande: Federal University of Paraiba, Campina Grande. 1999. 218p. (FAO Studies: Irrigation and Drainage, 29 Revised).

ALVARES, C. A.; STAPE, J. L.; SENTELHAS, P. C.; GONÇALVES, J. L. M.; SPAROVEK, G. Koppen'sclimate classification map for Brazil. Meteorologische Zeitschrift, v. 22, p. 711-728. 2013.

BADER, M. S. H. Seawater versus produced water in oil-fields water injection operations. Desalination, v. 208, n. 1-2, p. 159-168. 2007.

COSTA, A. G.Utilisation of dairy effluent in the production of thornless mandacaru (Cereus hildmannianus k. schum) in the Brazilian semi-arid region. 2017. 86p. Dissertation (Master's) - Universidade Federal Rural do Semi-árido, Mossoró, 2017a. Available at: <https://ppgmsa.ufersa.edu.br/wp-content/uploads/sites/59/2014/10/ DISSERTACAO-Andrezza-Grasielly-Costa.pdf>. Accessed on: 10 September 2018a.

COSTA, A. G; ALVES, S. M. C. ; BATISTA, R. O. ; MENDES, H. C. ; SOUZA, D. I.M. . Chemical Characteristics of Irrigated Ultisol with Dairy Effluent Dilutions. FLORAM, v. 25, p. 1-9, 2018.

CURBELO, F. D. S. Study of the removal of oil from water produced in the oil industry by column adsorption using expanded and hydrophobised vermiculite. 2002. 102 p. Dissertation (Master's) - Federal University of Rio Grande do Norte, Natal. 2002.

CRISÓSTOMO, L. A.; Weber, O. B.; MIRANDA, F. R.; ARAGÃO, F. A. S.; BEZERRA, F. C.; MOSCA, J. L. Reuse of produced water in ornamental pineapple irrigation: effects on soil chemical characteristics. Fortaleza: Embrapa Agroindústria Tropical, 2018. 27 p.

BRAZILIAN AGRICULTURAL RESEARCH COMPANY - EMBRAPA. System Brazilian soil classification. 5 ed. Brasília: Embrapa Solos; 2013. 590p.

FAKHU'L-RAZIA, A.; PENDASHTENA, A.; ABDULLAHA, L.; BIAK, D.; MADAENIC, S. & ABIDINA, Z. Review of technologies for oil and gas produced water treatment. Journal of Hazardous Materials. vol. 170,p. 530-551, Oct. 2009.

FERREIRA, B. H. Study of oil produced water treatment processes. 2016. 48 p. TCC (Graduation) - Petroleum Engineering Course, Federal University of Rio Grande do Norte, Natal, 2016. Available at: <https://monografias.ufrn.br/jspui/bitstream/123456789/3067/1/TCC-2016%202-%20BARBARA%20HELINSKA%20FERREIRA.pdf>. Accessed on: 10 January 2019.

GABARDO, I. T. Chemical and toxicological characterisation of produced water discharged from oil and gas platforms off the Brazilian coast and its dispersive behaviour at sea. 2007 250 p. Thesis (Doctorate) - Federal University of Rio Grande do Norte, Natal. 2007.

MIRANDA, F. R.; CRISOSTOMO, L. A.; WEBER, O. B.; SILVA, F. L. B.; ARAGÃO, F. A. S.; BARROS, M. E. S. Irrigation with water produced in oil extraction: effects on soil salinity and productivity of papaya. Fortaleza: Embrapa Agroindústria Tropical, 2016.29 p. Available at: <https://ainfo.cnptia.embrapa.br/digital/bitstream/item/147549/1/BPD16010.pdf>. Accessed on: 12 Nov. 2018.

MOTTA, A. R. P.; BORGES, C. P.; ARAUJO, K. P. & BRANCO, L. P. N. Treatment of petroleum produced water for oil removal by membrane separation processes: a review. Sanitary and Environmental Engineering, v. 18, p.15-26, 2013.

NEFF, J.; LEE, K.; DEBLOIS, E. M. Produced water: Overview of composition, fates and effects. Chap. 1 In: Produced water, environmental risks and advances in mitigation technologies. Springer Science, 608p. 2011.
NSW - Effluent reuse management. Australia: Department of Environment and Climate Change and Water NSW, 2010. 65 p.

OGP, 2005. Fates and effects of naturally occurring substances in produced water on the marine environment. International Association of Oil and Gas Producers. v. 364, 42p.

PINHEIRO, R.; TRAVALLONI, A. M.; JÚNIOR, G. M.; CRISÓSTOMO, L. A.;AQUINO, O.; ORLANDO, A. E. Pilot Irrigation Project with Produced Water in the Fazenda Belém Field. In: RIO OIL & GAS EXPO AND CONFERENCE, Rio de Janeiro. Proceedings... Rio de Janeiro. 2014. Available at: < https://ainfo.cnptia.embrapa.br/digital/bitstream/item/111139/1/AAC14001.pdf>. Accessed on: 30 January 2019.
RÊGO, L. G. S.; MARTINS, C. M.; SILVA, E. F.; SILVA, ALVES, J. J; LIMA, R. N.S. Pedogenesis and soil classification of an experimental farm in mossoró, state of Rio Grande do Norte, Brazil. Revista Caatinga, n. 29, p. 1036-1042, 2016.
RIBEIRO A.C, GUIMARÃES P. T. G, ALVAREZ V. H. Recomendações para o uso de corretivos e fertilizantes em Minas Gerais. Viçosa: CFSEMG; 1999. 322p.
SILVA, I. N; FONTES, L. O; TAVELLA, L. B; OLIVEIRA, J. B; OLIVEIRA, A. C;
Irrigation water quality. Agropecuária Científica no Semi-Árido, Campina Grande, v. 3, n. 7, p.01-15, jul. 2011.
TEIXEIRA, P. C.; DONAGEMMA G. K.; WENCESLAU, A. F.; TEIXEIRA, G.Manual of soil analysis methods. 3. ed. Rio de Janeiro: Embrapa Solos, 2017. 573 p.
WANG, X.; GOUAL, L.; COLBERG, P. J. S. Characterisation and treatment of dissolved organic matter from oilfield produced waters. Journal of Hazardous Materials,v. 30, p. 164-170. 2012.
VELOSO, M. E. C. Effects of different water qualities on some physical-chemical properties of a reddish Brunizem. 1991. 51p. Dissertation (Master's Degree) - Federal University of Viçosa, Viçosa, 1991. Available at: <http://www.scielo.br/scielo.php?script=sci_nlinks&ref=000177&pid=S0100-06832009000600031&lng=pt >. Accessed 2 Oct. 2018.

MIX
Papier aus verantwortungsvollen Quellen
Paper from responsible sources
FSC® C105338

Printed by Books on Demand GmbH, Norderstedt / Germany